# SCIENCE EXPERIMENTS

Experiments to spark curiosity and
develop scientific thinking

# Tricia Dearborn

Advisor: Trevor Davies

A & C Black • London

# CONTENTS

# INTRODUCTION

The best way to learn about science is to do it. Sparking children's scientific imaginations is as important as teaching them why an apple falls to the ground. *Science Experiments* offers a range of practical experiments which cover all the major strands of primary school science. Dip into it when you need an easy-to-organise activity to supplement a particular unit of work within your science teaching. These experiments will help children to develop observational skills and learn ways of hypothesising, testing and recording. They can all be carried out by the children themselves, under supervision.

## ABOUT THIS BOOK

### TEACHERS' FILE

The teachers' file offers advice on how to make the most of this book. There are background notes and suggestions for classroom organisation, assessment and parental involvement, as well as a handy list of basic materials. The 'Please Explain' section provides notes on all the experiments in the activity bank.

### QUICK STARTS

These activity ideas are a great way to grab children's attention at the beginning of a lesson. They require little or no preparation, and can be used in any order. There is a page of ideas for each science unit.

### ACTIVITY BANK

This section contains 24 photocopiable experiments covering Physical Processes, Life Processes and Living Things, Materials and Their Properties, and Earth and Space. These are followed by three activities which deal with working safely and scientifically. The activity sheets can be used in any sequence. Each experiment shows the relevant learning objective.

### CHALLENGES

Pupils can use these photocopiable task cards to help them design their own experiments, working individually, in pairs or in small groups. The cards can be used in any sequence. Each one includes a focus question followed by ideas for planning the experiment and presenting the results.

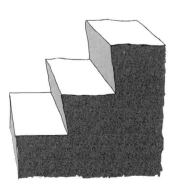

# HOW TO USE THIS BOOK

## QUICK STARTS

Quick Starts are ideal as warm-up activities at the beginning of the lesson. Each activity is intended to provide 10–15 minutes of group or whole class discussion. Reflect on the completed task with the children. Ask what they learned and whether there was anything that surprised them.

**Example** Sticky salt (page 16) is an ideal small group experiment. It will give children the opportunity to discuss their findings in their group before feeding back to the whole class.

### Sticky salt

Each child needs two saucers, black paper, salt and a pencil. The children should line each saucer with paper, and sprinkle some salt onto each. Ask them to hold one saucer about 15 cm from their mouth and breathe onto it steadily for about two minutes, then stir each pile of salt with the pencil. The water vapour in breath causes salt crystals to stick together. Similarly, the salt in salt shakers gets sticky when the weather is humid.

## ACTIVITY BANK

These photocopiable activities can be used by individuals, groups or the whole class, and can provide the focus for a whole lesson (most of the activities require 30–40 minutes of investigation). The investigation could then be followed up by ICT or library research. The objective of each activity is referenced to the National Curriculum.

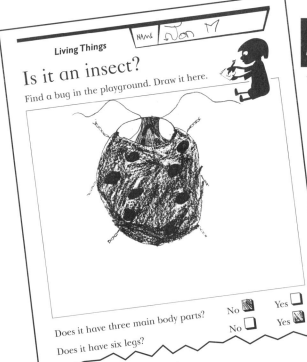

**Living Things**   NAME Jon M

### Is it an insect?

Find a bug in the playground. Draw it here.

Does it have three main body parts?  No ☒  Yes ☐

Does it have six legs?  No ☐  Yes ☒

**Example** Is it an insect? (page 28) provides an opportunity for children to observe an insect. It may be necessary to find the insects in advance, to ensure there are different kinds. Place each insect in a clear container so that each child can see all of it. Magnifying glasses would also be useful. When discussing the insects introduce scientific vocabulary as appropriate, and write useful words on the board.

## CHALLENGES

These photocopiable activities are perfect for use in learning centres, in the school library or in the classroom. The investigational nature of the activities is in line with the requirements of Sc1 Scientific enquiry. All of the Challenges support the development of investigation, enquiry and problem-solving skills.

**Example** What can you make into a magnet? (page 46) is a challenge that would be best done as a group activity. Encourage groups to write a hypothesis to answer the question first and test it in a scientific way. They should keep a record of their findings to use in their presentation.

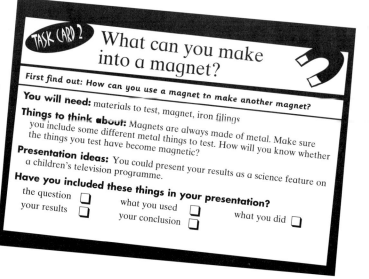

TASK CARD 2   **What can you make into a magnet?**

First find out: How can you use a magnet to make another magnet?

**You will need:** materials to test, magnet, iron filings

**Things to think about:** Magnets are always made of metal. Make sure you include some different metal things to test. How will you know whether the things you test have become magnetic?

**Presentation ideas:** You could present your results as a science feature on a children's television programme.

**Have you included these things in your presentation?**

the question ☐     what you used ☐
your results ☐     your conclusion ☐     what you did ☐

# TEACHERS' FILE

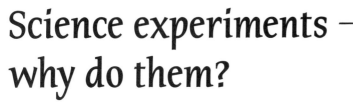

# BACKGROUND NOTES

## Science experiments – why do them?

Through doing science experiments at school, children not only have the opportunity to explore for themselves how the world works, they also become familiar with scientific thinking: how to ask a question, form a hypothesis, design a way to test it, gather results and form a conclusion.

The hands-on experimental approach has far greater impact than the simple learning of scientific facts from books. When children have made their own rainbows – seen for themselves white light entering a cup of water and different colours landing on the paper at the other side – it will be much easier for them to remember that white light is made up of light of different colours. What is more, in the process of doing these experiments pupils will develop many valuable skills, including those of observation, handling equipment, techniques for working with others, and consideration for safety in what they do.

The experimental approach enriches children's lives by encouraging them to be curious about the world around them. And in addition to all this, science experiments are fun!

## Motivation

At the end of most of the activity sheets, the children are asked to suggest an explanation for what happened. Encouraging pupils to think of possible explanations for a phenomenon is a vital step – it gives them a greater level of participation and leads them to be more enquiring. This also helps pupils learn how to form a hypothesis – a postulated reason for what happens, which can then be tested. If they can think of a way of testing their explanation, encourage them to carry it out.

Once the children have made their suggestions, explain the scientific principle behind the experiment. Clear explanations for each activity sheet can be found in the section 'Please Explain' on pages 10–11. If a child records a different result from the findings of the rest of the class, ask the child to repeat the experiment. It may be useful to ask another pupil to observe, just to check that the same procedure has been applied.

## Conducting their own experiments

Pupils can use the activity sheet on page 42 to help them design and carry out their own experiments. Encourage them first to write a question that they would like to answer (for example, 'Does a black crayon melt more quickly in the sun than a white crayon?'). Then ask them to propose a hypothesis (for example, 'I think the white crayon will melt more quickly because it is shinier.'). They should decide how they will test their hypothesis, and work out what equipment and materials they will need.

The children should also consider how to record the results, and what criteria they will be observing. The data record sheet on page 43 may be useful for this. Depending on the type of experiment, the results could be presented as a chart, a graph, or as a clear written account of what happened. The children could also use simple computer word-processing or graphing programs for recording and presenting results.

The results themselves need to be considered carefully. Ask the children whether the hypothesis seems to be correct, why they think this and what conclusion they can come to. Remind them that there is no shame in being completely wrong in their hypothesis! A hypothesis simply provides a starting point. What matters is how they respond to the information gathered in their experiment. Occasionally, the children may find their results inconclusive. Talk to them about why this might be. Ask them if they think it might be useful to try the experiment again, and whether to first modify something in the design of the experiment.

# CLASSROOM ORGANISATION

# General

Science experiments provide opportunities for the children to work individually, in pairs or in groups. Some of the experiments will need to be done outdoors, while others can be done by pupils at their desks.

It can be useful to have a science scrapbook for each pupil. Encourage the children to glue in their completed activity sheets, along with any other relevant materials. This will enable them to keep all their science records together, and to review their progress over the year by looking through the scrapbook.

# A Science Open Day

The task cards in the 'Challenges' section of this book can be used as starting points for displays for a Science Open Day. Displays could also be generated from most of the experiments in the activity bank. Invite parents and other classes to view the displays. Presentations could be visual – for example posters or labelled models – or oral. Displays could also be set up so that visitors to the Open Day can perform the experiment (or part of it) themselves. Refreshments could be provided on a scientific theme.

Encourage the children to use different media for their displays, when appropriate. For example, a pupil could show a video of an experiment, or play appropriate background music. However, additional media should only be used if they enhance the presentation or add to its clarity. Sometimes the simplest displays are the best.

Take photographs at the Open Day – or arrange for a parent volunteer to do so – and make a display board showing children with their projects. The children will enjoy seeing themselves with their work, and it is a good way to celebrate all the hard work that went into organising the Open Day.

# Safety in the science classroom

It is important to talk to the class about general safety before doing science experiments. Photocopy a list of safety rules (see page 44) and display it on the classroom wall. The children could also come up with safety ideas of their own, which could be written on a separate sheet and displayed next to the original list.

Some of the experiments in this book involve tasting non-harmful materials. Make sure children understand that they should *never* taste anything in a science experiment unless their teacher tells them it is safe to do so.

# Setting up a science centre

A science centre can easily be set up in your classroom. A small table or desk should provide adequate workspace. Add a sign and some scientific decorations. It's not necessary to keep all the scientific equipment and materials nearby – just have out the materials children will need for the current experiment. Next to the science centre, keep a rubbish bin, a bottle of water, some cloths or paper towels for wiping up spills, and a bucket (for liquid experiments to be tipped into; empty it at the end of the day).

Decide whether the children will do the experiments individually, in pairs or in small groups. Read the activity sheet with the whole class first, and demonstrate any specific procedures that are new or might be difficult. Ask the children to tick off the materials on the sheet as they identify them, and tick off each step as they complete it. While the children are using the science centre, check that they are following the correct procedures, writing down their observations and working safely. Make sure they understand the importance of cleaning up when they have finished and leaving the science centre how they found it.

The equipment for each experiment can be set up and left for a week or two. Change experiments after two weeks at the most, so that the children still remember the experiment when they come together as a class to discuss their results and observations. This discussion time gives children the opportunity to compare results and consider possible reasons for discrepancies.

# THE BASIC EQUIPMENT

## Materials

aluminium foil*
cooking oil
cornflour
egg cartons
eggshell pieces
food colouring (red and blue)
flooring materials (lino, boards,
    carpet squares, doormat)
iron filings
lollipop sticks
masking tape
modelling clay*
paper towels or kitchen paper
plastic (supple, black and white)
plastic straws*
rubber bands
salt
sand
skimmed milk powder
string
sugar
tissue boxes (empty)
vinegar

You may also need one-off items for particular experiments – for example, lemons for 'Electric lemons' on page 13, or the pot plants used in the experiment on page 26.

## The science cupboard

This page gives lists of materials and equipment that the children will need for the experiments in this book. Those marked with an asterisk (*) are not required for these experiments, but are useful to have on hand. Most of the items listed are either household materials or easy to obtain at educational toy shops or supply companies.

Make sure the science materials cupboard is lockable, and kept locked, if you keep anything in it that could be harmful to the children.

## Equipment

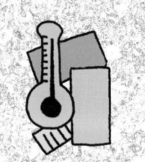

balloons
bar magnets
blindfolds
bowls (sets of different sizes)
buckets
copper wire (insulated)
cups (clear plastic)
eye droppers
funnels*
glass jars (some should have lids)
ice-cube trays
magnets
magnifying glasses
matches
needles
paperclips
spoons (sets of different sizes)
stopwatches*
thermometers (class set)
torch batteries
torches

## Care of materials and equipment

Ensure that these basic tips are followed for storing materials and using equipment:

- Keep dry materials in airtight containers.
- Measure out all dry materials with a clean implement so that other materials don't get mixed into them.
- All equipment should be cleaned after each experiment, if necessary, and put away where it belongs.
- Always use alcohol thermometers rather than mercury ones. Mercury is highly poisonous and very difficult to dispose of in the event of breakages.

# TECHNOLOGY

Information and communication technology provides valuable tools for reporting scientific experiments. Desktop publishing programs designed for younger pupils can be used for presenting experiment results. Programs which feature database and graphing facilities are also available, as are age-appropriate CD-ROM encyclopaedias.

The Internet is a valuable resource: scientific topics, discoveries, inventions and scientists' names can be typed into search engines to find relevant websites. (The children may need assistance to make their searches specific enough to be useful.) When it comes to whole class presentations such as a science fair, encourage children to use different media for their displays. They could include audio or video presentations, background music on CD, or use desktop publishing to produce pamphlets for visitors to take away.

# ASSESSMENT

Monitoring the way children conduct and report scientific experiments will enable you to assess a number of factors, among them pupils' understanding of scientific concepts, proficiency in the use of scientific equipment, development of observational skills, and ability to draw conclusions from their results.

Assessment of experiments should be based on how the children go about it, rather than on whether or not they get the 'right' answer. For example, you can note whether each pupil has:

- listened to/taken part in the initial discussion of the experiment
- followed the procedure accurately and with regard to safety
- co-operated with classmates if working in a group
- kept accurate notes
- presented results
- stated a conclusion and given reasons for it.

Once the children begin to devise their own experiments, they can also be assessed on their ability to form a clear question as a starting point.

Completed activity sheets can be kept in the pupils' portfolios as a record of progress. Encourage children to select their own samples for the portfolio to help them develop skills in self-assessment.

# PARENTAL INVOLVEMENT

If any of your pupils have parents who are scientists, ask if they would be willing to come into the classroom to talk to the class about what they do, what kind of science is involved, and how they became interested in it in the first place. If their job involves scientific procedures which could be safely demonstrated to the class, this would add interest to the talk.

Parent volunteers can be called on to help collect or prepare materials for the science centre. Giving parents the opportunity to see the science centre in action is a good way of involving them in the class's science programme. A Science Open Day can also be a great way to involve parents (see page 7 for ideas).

Parents may also be willing to accompany the class on science-based excursions, for example to nature reserves, observatories, museums or exhibitions.

Encourage the children to take their activity sheets home to show their parents. Many of the experiments can be continued at home, for example weather watching, plant classification, observation of shadows, identification of insects. If you have any materials which may help parents to spot opportunities for discussing science in everyday situations, give them to the children to take home.

Also keep an eye out for interesting and appropriate science programmes on television. These can be watched at home and discussed with parents, as well as talked about in the classroom.

**Page 18 Make your own rainbow!** White light is made up of light of different colours. When light passes at a slant from the air into the glass of water, the rays of light bend. Each colour bends at a slightly different angle, which means the different colours travel in slightly different directions, and land on the paper in different places. The same thing happens with a rainbow in the sky when the sun shines at an angle through water drops in the air.

**Page 19 The colours of sunset** White light is made up of light of different colours. The particles of milk filter out some of the colours, allowing only red and orange light to shine through to the wall. In a sunset, smoke and dust particles filter the sunlight, and the sky turns red and orange.

**Page 20 Melting moments** The black plastic looks black because it *absorbs* all the colours that make up white light. The light energy is converted to heat, which melts the ice cube. The white plastic looks white because it *reflects* all the colours that make up white light. It absorbs a lot less light energy, and the ice cube melts more slowly.

**Page 21 Rubber band rock** If the tissue boxes have plastic in the hole, cut it out before giving them to the children. Use medium rubber bands rather than narrow ones. The sound is caused by the rubber bands vibrating. All sound is caused by vibrations. The sticks lift the rubber bands up and give them more room to vibrate. The tissue box acts like the sound box of a guitar, amplifying the sounds (making them louder).

**Page 22 Amazing magnets** This experiment will work best if you use the strongest magnets you have, and square or rectangular containers with clear bottoms. When the mixture in the container is placed on or near a magnet, the iron filings become magnetised. They line up within the magnet's *magnetic field*. A magnetic field is the space near a magnet through which a magnetic force acts. The filings line up along what are called the *lines of force*.

**Page 23 Bouncing balloons** Some ideas for balloon propulsion: uninflated balloons could be thrown, kicked, stretched back like a catapult and released; inflated balloons could be hit, blown, or released without being sealed so that the air inside propels them.

**Page 24 Is it alive?** Living things need food. They grow, and some are capable of moving. How can the children tell a plant is alive? (It grows, and it needs food from the soil. Some plants also move, e.g. to follow the sun.) What about a river, which moves and can grow bigger after rain?

**Page 25 Thirsty celery** If you hold a tiny tube upright in water, the water moves up inside it because the water is attracted to the sides of the tube. This is called capillary action. The stem of a plant contains lots of tiny tubes. Water absorbed by the roots moves along the tubes up the stem of the plant. The coloured water moves up the tubes in the celery stalk and colours the leaves.

**Page 26 A place in the sun** The plants both have the same soil, the same food reserves, and as much water as they need. The only difference is that one plant receives sunlight and one does not. The results must be due to the factor that was different; in this case, the amount of sunlight the plants received.

**Page 27 Do seeds need water to grow?** The seeds in the jars where water has been added should germinate – that is, sprout little roots and shoots. Seeds need water to grow, so unless there is water there for them, they usually will not germinate.

**Page 28 Is it an insect?** Make sure that the children do not touch or harm their minibeasts. Insects have three main body parts and six legs. Many also have two pairs of wings. Spiders have two main body parts and eight legs.

**Page 29 Ant picnic** You could use fruit, meat, sugar, flour, honey, lettuce. Choose an ants' nest away from school buildings and make sure that the children don't walk on the nest. Foraging ants will usually look for sugars and fats to take back to the colony, but will take proteins back to the queen ant when she is laying a lot of eggs.

**Page 30 Fizzy stones** Ask the children to collect interesting-looking stones from the playground. If the stones fizz, or bubbles appear, they probably contain limestone. Limestone contains calcium carbonate, and this reacts with the vinegar (which is a weak acid) to form carbon dioxide gas, which bubbles to the surface. Eggshells also contain calcium carbonate.

**Page 31 The taste test** A lot of what we think of as our sense of taste is really our sense of smell. The tongue can detect tastes which are sweet, salty, sour (like a lemon) and bitter (like a grapefruit). More complex flavours are detected by the nose. Foods which have similar textures can be hard to tell apart when your nose is blocked.

**Page 32 Does it absorb?** Provide samples of absorbent materials (e.g. kitchen paper, terry towelling), and non-absorbent materials (e.g. plastic, greaseproof paper). When the children have finished, discuss how a material's absorbency or non-absorbency can make it useful in particular situations. Ask them, 'What would happen if raincoats were made of terry towelling? Or if hankies were made of plastic?'

**Page 33 Growing crystals** The children will need individual supervision when stirring sugar into the hot water. Emphasise that they should *never* taste anything used in a science experiment unless the teacher tells them it is safe. The sugar dissolves in the hot water. It is still in the water (the children can check this by tasting it once it has cooled) but you cannot see it. When the water evaporates from the solution, the sugar remains.

**Page 34 Making rain** Some of the water in the jar *evaporates* – turns into water vapour, which is an invisible gas. When the water vapour hits the cold lid, some of the vapour *condenses* – turns back into liquid water. When there is enough water on the bottom of the lid, it drips back into the jar. Evaporation, condensation and melting are examples of reversible changes. The water is still water, but changes from one form to another (e.g. from a solid to a liquid, or a liquid to a gas) and it can also change back again.

**Page 35 Oil and water** Oil floats on top of water because it is less dense than water. If you have the same volume of oil as of water, the oil weighs less, e.g. a cup of oil weighs less than a cup of water. Oil and water do not mix at all. The scientific term for this is *immiscible*.

**Page 36 Shifty shadows** A shadow is formed when an object blocks the light coming from a light source. A shadow is the absence of light.

**Page 37 Do shadows move?** To make a very simple sundial, the children could mark the shadow's position each hour, on the hour, for one full day.

**Page 38 The weather today** Ask the class to record the weather at the same time each day. Add extra symbols if you need them, e.g. if you are likely to have snow. You could continue for another week,

predicting the following day's weather and recording both the prediction and the actual weather. Ask the children if they notice any patterns in the weather.

**Page 39 Stop that soil!** The grass slows the flow of water across the surface of the soil and the roots of the grass hold the soil together, so less soil should be washed down that tin. When soil is washed away by water or wind, it is called erosion. When land has been over-farmed so that most of the grass is eaten, erosion can become a serious problem.

**Page 40 The mysterious vanishing wate**r Liquid water molecules on the surface of the water absorb energy from the surrounding air and change into water vapour. If the jar is sealed, the water vapour cannot go anywhere, and it condenses (turns back into liquid water) when it hits the cool surface of the jar. If the jar has no lid, the vapour escapes into the atmosphere, and the water level in the jar goes down. This is what happens to lakes and ponds if the water that evaporates is not replaced by rain.

**Page 41 Drip, drip, drop!** Water molecules are attracted to each other. When drops of water are pushed together, they join to form a larger drop. When the drop is large enough, it will fall to the ground. This is what happens in clouds – tiny water droplets in the air join to form larger, heavier ones which fall as rain.

**Page 42 My experiment** Make sure that the children begin with a clear, simple question. 'What I think will happen' is their hypothesis. 'What I will need' and 'What I will do' gives the method. Discuss with the children how they will record and present the results of their experiment. In the conclusion, they should consider whether their results support their hypothesis or contradict it, and what they think might be the reasons for this. If the results are inconclusive, the child may wish to repeat the experiment. Go through the procedure first with the child, in case modification of the method is necessary.

**Page 43 Data table** This handy table can be used to record results for many experiments. Work with the children to decide on headings for the rows and columns.

**Page 44 Safety in the science classroom** Photocopy and put up in a prominent place. Discuss safety issues with the class and encourage children to brainstorm their own list of safety rules. Add to the list any that you and the class feel are useful.

QUICK STARTS

# Physical Processes

## A bendy mirror

Ask the children to hold up a spoon as if it were a mirror and look into the convex side (which bends outwards). What do they see? Then turn it around and look at the concave side (which bends inwards). A flat mirror reflects light straight back to your eye. Because the spoon is curved, the light travels different distances to get back to your eye. What you see is distorted, and, if the mirror is concave, the image turns upside down!

## The magic egg

Grease the opening of a bottle with margarine, crumple a small piece of paper, light it and drop it into the bottle. Quickly place a peeled, hard-boiled egg on top. (The bottle opening should be slightly smaller than the egg.) The burning paper uses up the oxygen in the bottle. The air pressure inside the bottle is then less than the air pressure outside, and the outside pressure pushes the egg into the bottle.

## Seeing colour ghosts!

Ask the children to colour a 10 cm square green, then stare at it for one minute before moving their eyes quickly to a white wall. What do they see? Cone cells in the back of your eye detect red, blue and green. When you stare for a long time at something green, the cones that see green get tired and when you look away you see a mixture of red and blue instead.

## Make your own magnet

Ask the children first to test a needle by putting it close to some iron filings. What happens? Then ask them to stroke a magnet along the needle in the same direction for about a minute. What happens when they put the needle near the iron filings now? Stroking the needle with the magnet makes atoms in the needle line up all the same way, which makes it magnetic too.

## Electric lemons

Show the children how to remove the insulation from each end of a piece of copper wire and insert one end into a lemon; then straighten one end of a paperclip and insert it into the lemon as well. Ask them to touch the ends of the paperclip and the wire to their tongues at the same time. The lemon acts like an electric battery, using two metals and an acid (citric acid) to create an electric charge.

## The oldest snail

This is a good activity for when it has just rained. Go outside with the class and look for snails. Take magnifying glasses so that you can look closely at the snails' shells. Ask the children to count the 'rings' on the snails' shells. Snails are molluscs, and add new 'rings' as they grow. Who can find the oldest snail? The youngest? Remind the children to be gentle with the snails, and to examine them without picking them up if possible.

## Pick your part!

Play 20 Questions, with one child choosing to 'be' any visible part of the human body, and the rest of the class guessing what it is. Example questions could be 'Is there only one of you?', 'Can you smell?', 'Are you bigger than a mouth?', 'Do I use you to walk?' The child who guesses the answer chooses the next part, and if no one guesses, the first child goes again.

## What kind of plant?

Take the class into the playground and ask them to find as many different plants as they can. Work with them to sort the plants into groups. Encourage them to pay special attention to the criteria they are using. These should be based on observable characteristics of the plants, such as size, colour, leaf shape, whether it has flowers.

## That special relationship

Discuss with the class the different relationships between living things. Sometimes this relationship helps them both (e.g. bees pollinating flowers), sometimes only one creature benefits (e.g. parasites), and sometimes one animal or plant is food for another. As a class, brainstorm as many kinds of relationships as you can think of. The children could form pairs and choose a relationship to act out, for the rest of the class to guess.

## The wonderful web

Ask one child to choose a living thing to be – for example, 'I'm a cow!' – and act it out. Can someone else think of (and be) another living thing that has a relationship to the first? For example, 'I'm the grass the cow eats.' 'I'm the human who milks the cow.' 'I'm the dog that lives with the human.' 'I'm the flea that lives on the dog.' Can the whole class be included to build one big web of relationships?

# Materials and Their Properties

## The magical substance

Cornflour doesn't dissolve in water, but forms a suspension with interesting properties. Mix up cornflour in water so that it is wet enough to flow, but dry enough to break off in pieces. Give the children a portion in a cup, and ask them to explore its properties. Is it a solid? Is it a liquid? What happens when they pour it? What happens when they poke or stir it? Add water if the mixture gets too dry.

## What food is that?

Assemble a range of food samples with something in common, e.g. the foods could all be a similar colour (pieces of orange, carrot, pumpkin) or texture (potato, apple, pear). Ask the children to work out what each sample is, and to note how they do this. Which senses are they using? Try it again with the children blindfolded, or holding their noses. What do they notice now?

## The barefoot test

Lay on the floor in a line a selection of materials, e.g. a square of carpet, a doormat, lino, boards, pebbles, sand. Let a group of children walk on them, first wearing shoes, then in bare feet. Ask questions, for example: What do you notice about each material? Which would work best to wipe dirt off shoes/as a non-slippery floor covering? Which would wipe clean most easily? Which are comfy for bare feet and would make good indoor floor coverings?

## A closer look

Ask the children to cut a hole 2.5 cm square in a piece of paper or card. They can place this over something they want to examine in more detail, using the hole as a frame. This technique helps children to focus on the fine detail of the material they are examining, and can be very useful when comparing the observable features of different materials.

## How things pour

Assemble some substances made up of crystals, e.g. sugar, salt, sand. Ask the children to pour them from one hand to the other. What do they notice about the way each substance pours? Let them examine the crystals through a magnifying glass. What do they notice about the shape and the size of the individual crystals? What effect do they think this would have on the properties of each substance?

# Earth and Space

## Sticky salt

Each child needs two saucers, black paper, salt and a pencil. The children should line each saucer with paper, and sprinkle some salt onto each. Ask them to hold one saucer about 15 cm from their mouth and breathe onto it steadily for about two minutes, then stir each pile of salt with the pencil. The water vapour in breath causes salt crystals to stick together. Similarly, the salt in salt shakers gets sticky when the weather is humid.

## A windy day

How windy is it? Look for signs like leaves and dirt blowing, trees bending. Discuss how a wind-sock shows how strong the wind is. Think of other materials that could be used like this, e.g. strips of paper or cloth, an actual sock. Attach different materials to the top of a fence, in order from most to least easily moved. Check the wind gauge on different days, noting the conditions that make each item move. The class can devise their own scale, e.g. a paper-lifting breeze, a sock-waggling gale.

## How long is a shadow?

Do this activity outside on a sunny day which looks like it will stay sunny. Ask the children to form pairs. One child stands still while their partner marks the spot where they are standing and the place their shadow ends. This should be repeated (e.g. hourly) through the day. The child whose shadow is being marked should stand in the same place each time. What happens to the shadows?

## Cold, cool, warm, hot

Set up four jars containing water, one labelled 'hot' (about 40°C), one 'warm' (25°C), one 'cool' (10°C), and one 'cold' (0°C). Ask the children to test the water with a finger before using a thermometer. What happens to the red line on the thermometer when they put it into the different jars? Once children are familiar with how a thermometer works, let them find things to test and guess the temperature by touch before measuring it.

## Weather predictions

For a week, observe and record the weather at the same time each morning. Ask the children to predict what the weather will be like tomorrow, and record their predictions. They could take into account the season, today's weather, particular signs (e.g. pinkish clouds can herald hail or snow). Do the children notice any patterns, or correspondences between particular observations and subsequent weather?

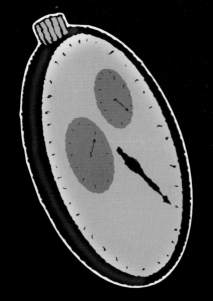

# ACTIVITY
# BANK

NAME

# Make your own rainbow!

**You will need:**

- a clear plastic cup
- tap water
- a sunny windowsill
- crayons, coloured pencils or felt-tipped pens

**What to do:**

1. Put some water in the cup.
2. Put the cup on the windowsill so that the sun shines through it.
3. Put this sheet of paper on the floor where the sunlight falls. To draw your rainbow, colour over each part of the rainbow with a colour that matches it.

*Objective: To observe that white light is made up of different colours*

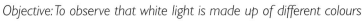

NAME

# The colours of sunset

## You will need:

- a clear plastic cup
- a white wall
- torch
- water
- spoon
- 1 tablespoon of skimmed milk powder

## What to do:

1. Put water in the cup. Leave the top 5 cm empty.
2. Put the cup in front of the white wall.
3. Shine the torch through the water towards the wall. What do you see?

_____

4. Stir the milk into the water.
5. Shine the torch through the milky water. What do you see?

_____

_____

_____

Why do you think this happens?

_____

_____

_____

*Objective: To observe that white light is made up of different colours*

NAME

# Melting moments

### You will need:

- 2 ice cubes
- a piece of white plastic
- a piece of black plastic
- a sunny day

### What to do:

1. Wrap one ice cube in the piece of white plastic.
2. Wrap one ice cube in the piece of black plastic.
3. Put both of them in the sun.
4. Check the ice cubes every two minutes. Record what you notice.

| Time (minutes) | Ice cube in white | Ice cube in black |
|---|---|---|
| 2 | | |
| 4 | | |
| 6 | | |
| 8 | | |
| 10 | | |

Write a sentence to explain what happened.

_____

_____

*Objective: To observe that light colours reflect light energy and dark colours absorb light energy*

NAME

# Rubber band rock

## You will need:

- an old tissue box
- 3 rubber bands
- 2 sticks or pieces of wood about 12 cm long

## What to do:

1. Put the rubber bands around the tissue box lengthways. Make sure they go over the hole.
2. Tuck the sticks under the rubber bands widthways, near the edges of the box.
3. Pluck the rubber bands with your fingers. What happens?

_____

_____

Press on the string with one finger while you pluck it with another. What happens?

_____

_____

Can you play a tune on your rubber band instrument?

*Objective: To observe that all sound is caused by vibrations*

NAME

# Amazing magnets

## You will need:

- 2 bar magnets
- half a cup of golden syrup
- 1 dessert spoon full of iron filings
- a spoon
- a clear plastic container

## What to do:

1. Mix the iron filings into the golden syrup with the spoon.
2. Pour the mixture into the plastic container.
3. Place the two magnets end to end. Leave a bit of space between them.
4. Place the container gently on top of the magnets.
5. Watch what happens.

Draw what you can see.

Why do you think this happens?

*Objective: To observe that a magnetic field is a space near a magnet through which a magnetic force acts*

NAME

# Bouncing balloons

How many different ways can you make a balloon move? List the ways. The balloon can be flat or blown up.

1. _____

2. _____

3. _____

4. _____

5. _____

6. _____

Which way moves the balloon the furthest?

_____

Why do you think that is?

_____

_____

_____

*Objective: To suggest questions about ways in which different objects move*

NAME

# Is it alive?

Choose three different things to test. You could choose a classmate, a ruler and a plant. Observe each one, then answer the questions. Write the answers in the chart below.

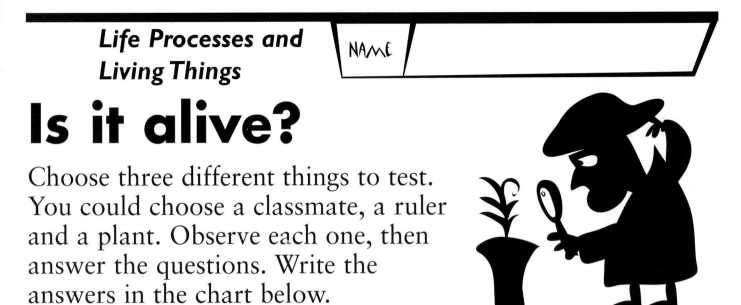

| **Name of item** | | | |
|---|---|---|---|
| **Does it move?** | | | |
| **Does it grow?** | | | |
| **Does it need food?** | | | |
| **Is it alive?** | | | |

*Objective: To observe that living things have special characteristics which show that they are alive*

# Thirsty celery

## You will need:

- 1 stalk of celery with leaves
- 2 tall jars
- water
- cup
- red food colouring
- blue food colouring
- teaspoon

## What to do:

1. Put half a cup of water in the bottom of each jar.
2. Add a teaspoon of red colouring to one jar. Add a teaspoon of blue food colouring to the other jar.
3. Ask your teacher to cut the celery lengthways along the stalk, from the bottom to just below the leaves.
4. Put one half of the stalk in the blue jar, and the other half in the red jar.
5. Let it sit overnight.

Draw a picture to show what has happened.

Why do you think this happened?

_____

_____

_____

_____

_____

_____

_____

_____

*Objective: To observe that water absorbed by roots moves upwards through the stem of a plant*

# A place in the sun

## You will need:

- 2 identical pot plants
- water

## What to do:

1. Put one plant in a cupboard, or in a box you can close.
2. Put the other plant in a sunny place.
3. Water each plant whenever the soil feels dry.
4. After a week, look at the two plants. What do you see? Record your results in the table.

|        | Plant in dark place | Plant in sunny place |
|--------|---------------------|----------------------|
| Size   |                     |                      |
| Colour |                     |                      |
| Leaves |                     |                      |
| Stem   |                     |                      |

Now put both plants in the sunny place. How long does it take for both plants to look healthy? _____

*Objective: To observe that green plants need light to grow*

Life Processes and Living Things

NAME

# Do seeds need water to grow?

**You will need:**

- 2 jars
- dried beans
- kitchen paper
- water

**What to do:**

1. Line each jar with some kitchen paper.
2. Add a little water to one of the jars.
3. Put some dried beans in each jar, between the paper and the glass.
4. Put the jars somewhere warm and dark for a week. What happens? Why?

_____

_____

_____

_____

_____

If any of the beans have little roots and shoots, put that jar in a light place. Check every day and watch what happens!

*Objective: To observe that plants need water to grow*

NAME

# Is it an insect?

Find a minibeast.
Draw it here.

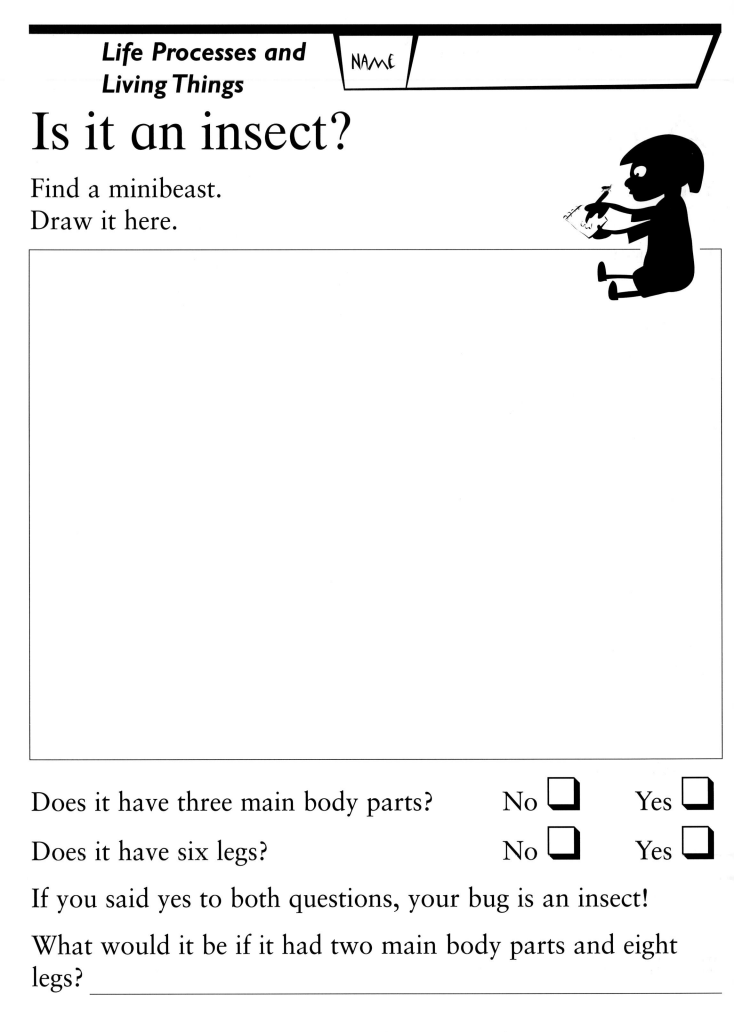

Does it have three main body parts?     No ☐     Yes ☐

Does it have six legs?     No ☐     Yes ☐

If you said yes to both questions, your bug is an insect!

What would it be if it had two main body parts and eight legs? _____

*Objective: To observe and identify the characteristics of an insect*

NAME

# Ant picnic

**You will need:**

- paper plate
- marker
- different foods

**What to do:**

1. Divide the plate into six sections, using the marker.
2. Put a piece of food in each section.

Which food do you think ants will like best? Why?

_____

3. Put the plate outside. Leave it there for a few hours.
4. Go back to look at the plate. What has happened?

| Food | Ants like it (✓) | Ants don't like it (✓) |
|---|---|---|
|  |  |  |
|  |  |  |
|  |  |  |
|  |  |  |
|  |  |  |
|  |  |  |

Which food did the ants like best? Why might this be?

*Objective: To observe how ants are attracted to different foods and why*

# Fizzy stones

## You will need:

- a glass bowl
- vinegar
- different kinds of stones and bits of eggshell

## What to do:

1. Pour some vinegar into the bowl.
2. Add the objects one at a time. Watch to see what happens.

| Object | What happens |
|--------|--------------|
|        |              |
|        |              |
|        |              |
|        |              |
|        |              |
|        |              |

Why do you think this happens?

_____

_____

_____

*Objective: To explore ways of grouping rocks and natural materials*

NAME

# The taste test

## You will need:

- a partner
- 4 small pieces each of apple, pear and carrot
- a blindfold

## What to do:

1. Blindfold the taster.
2. Give the taster a piece of apple. If they can tell what it is, put a tick in the grid. Put a cross if they can't. Try it with the carrot, then with the pear.
3. Repeat, but this time ask the taster to hold their nose.
4. Swap places and do it all again.

|  | **Apple** | **Carrot** | **Pear** |
|---|---|---|---|
| First taster |  |  |  |
| With nose held |  |  |  |
| Second taster |  |  |  |
| With nose held |  |  |  |

Write a sentence to explain your results.

_Objective: To observe that we have five senses which allow us to find out about the world_

# Does it absorb?

**You will need:**

- egg carton
- eye dropper
- water
- pieces of different materials

**What to do:**

1. Put the materials into different sections of the egg carton.
2. Use the eye dropper to squeeze a few drops of water onto each piece. Watch what happens.

Which materials absorbed water?

_____

_____

Which materials didn't absorb water?

_____

Which might you use to mop up a spill in the kitchen?

_____

Which might you use to make a raincoat?

_____

*Objective: To test whether materials are absorbent*

# Growing crystals

## You will need:

- a jar of hot water
- sugar
- teaspoon
- string
- lollipop stick

## What to do:

1. Carefully stir a teaspoon of sugar into your jar of water. Stir until all the sugar dissolves.
2. Stir more sugar in, a spoonful at a time. Do this until some sugar sits on the bottom no matter how long you keep stirring.
3. Tie the string around the lollipop stick.
4. Put the stick across the jar so that the string dangles in the liquid.
5. Put the jar in a safe place. Check it every two or three days. When all the water has gone, what do you see?

_____

_____

What do you think has happened?

_____

_____

*Objective: To explore the process of dissolving*

# Making rain

**You will need:**

- a jar with a lid
- water
- 3 or 4 ice cubes

**What to do:**

1. Put some water in the bottom of the jar.
2. Turn the lid upside down. Put it on top of the jar.
3. Put the ice cubes in the lid.
4. Watch the bottom of the lid for 10 minutes. Draw what you see.

| At first | After 5 minutes | After 10 minutes |
|---|---|---|
|  |  |  |

Write a sentence to explain this.

_____

_____

_____

*Objective: To observe the process of evaporation and condensation*

# Oil and water

## You will need:

- a small jar with a tight lid
- water
- blue food colouring
- spoon
- cooking oil

## What to do:

1. Fill the jar one-third full with water.
2. Add two drops of food colouring. Stir.
3. Very slowly fill the next third of the jar with oil.
4. Screw the lid on tightly.
5. Shake well.
6. Let the jar stand for three minutes.

Draw what you see.

Why do you think this happens?

_Objective: To observe the properties of different liquids_

NAME

# Shifty shadows

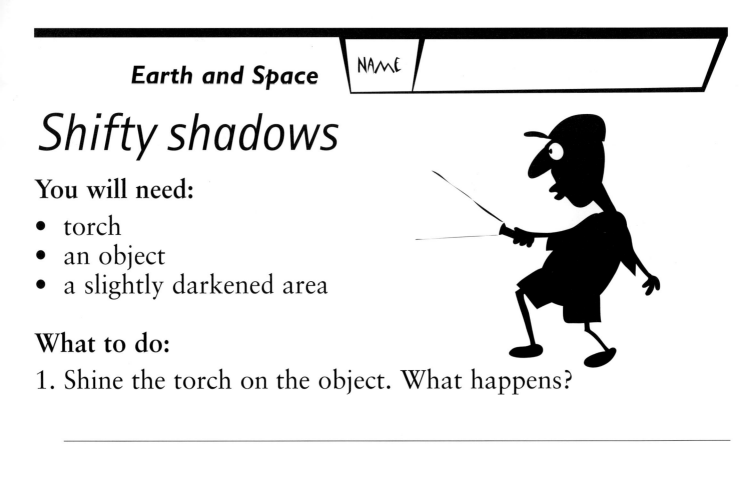

## You will need:

- torch
- an object
- a slightly darkened area

## What to do:

1. Shine the torch on the object. What happens?

_____

_____

_____

2. Move the torch further away from the object. What happens?

_____

_____

_____

What do you think a shadow is?

_____

_____

_____

*Objective: To observe that a shadow is the absence of light*

NAME

# Do shadows move?

**You will need:**

- pencil
- felt-tipped pen
- a sunny day

**What to do:**

1. Push the sharp end of a pencil through the dot in the middle of this sheet.
2. Push the pencil into the ground.
3. Using a felt-tipped pen, make a mark where the shadow of the pencil crosses the circle. Write next to it what time it is.
4. After an hour, mark the shadow again. Has it moved?

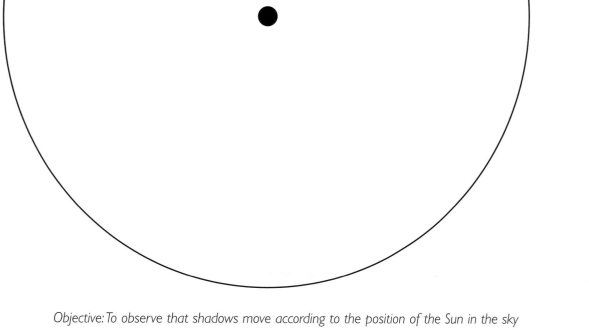

*Objective: To observe that shadows move according to the position of the Sun in the sky*

NAME

# The weather today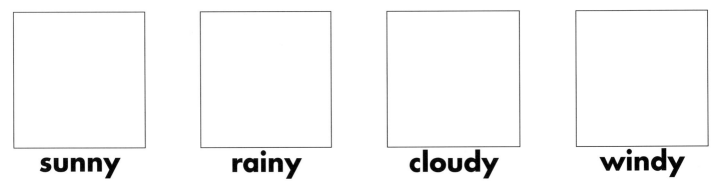

Design your own symbols for different kinds of weather.
Draw them in the boxes.

| | | | |
|---|---|---|---|
| | | | |
| **sunny** | **rainy** | **cloudy** | **windy** |

Record the weather each day from Monday to Thursday. Use
your symbols. You can use more than one. Try to predict
what the weather will be like on Friday.

| | |
|---|---|
| Monday | |
| Tuesday | |
| Wednesday | |
| Thursday | |
| My prediction for Friday: | |

*Objective: To observe that weather changes due to atmospheric forces*

# Stop that soil!

## You will need:

- 2 cake tins
- soil
- grass seeds
- blocks of wood
- water
- jug

## What to do:

1. Fill the tins with soil.
2. Plant grass seeds in one tin.
3. Water the soil in both tins equally.
4. Put the tin with seeds in it in a sunny place. Water it for several days.
5. When the grass is about 2 cm high, prop up one end of each tin on a block of wood so that the tins are at a slant.

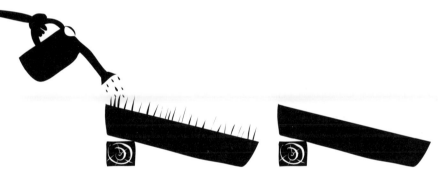

6. Pour water from the jug into the higher end of each tin. What happens?

Tin with grass: _____

Tin without grass: _____

Why do you think this happens?

_____

_____

*Objective: To observe how plants are important in preventing soil erosion*

# The mysterious vanishing water

**You will need:**

- 2 jars
- a lid for one of the jars
- water
- masking tape
- marker

**What to do:**

1. Put a strip of masking tape down the side of each jar.
2. Fill each jar half full with water.
3. Mark the level of the water on the masking tape, using a marker.
4. Put the lid on one jar.
5. Leave the jars for a week. Mark the new water levels.
6. Leave the jars for another week. Mark the new water levels.

What happened?

_____

_____

Why do you think this happened?

_____

_____

_____

*Objective: To observe that water evaporates when open to the atmosphere*

NAME

# Drip, drip, drop!

**You will need:**

- eye dropper
- water
- a see-through plastic lid
- pencil

**What to do:**

1. Fill the eye dropper with water.
2. Hold the plastic lid upside down.
   Squeeze onto it as many drops as you can fit.
3. Quickly turn the lid over.
4. Use the point of the pencil to move the drops of water together.

What happens. Why?

_____

_____

_____

_____

_____

_____

_____

_____

*Objective: To observe that water molecules are attracted to each other*

NAME

# My experiment

My question: _____

_____

What I think will happen: _____

_____

What I will need: _____

_____

What I will do: _____

_____

The results:

_____

My conclusion: _____

_____

# Data table

| | | | | | |
|---|---|---|---|---|---|
| | | | | | |
| | | | | | |
| | | | | | |
| | | | | | |
| | | | | | |
| | | | | | |
| | | | | | |
| | | | | | |
| | | | | | |

# *Safety in the science classroom*

Never touch, taste or smell anything unless your teacher says you can.

Ask for help if you need to cut anything.

Ask for help if you need to heat anything.

Clean up when you have finished.

# CHALLENGES

## TASK CARD 1
# What materials can light travel through?

**You will need:** different materials to test, a light source

**Things to think about:** Try as many different materials as you can. Can you devise a scale to rate how much light a material lets through?

**Presentation ideas:** You could stick pieces of the materials you tested onto a poster. Or as part of your display, you could set up the equipment as you used it, and let other people test for themselves.

**Have you included these things in your presentation?**

the question ☐         what you used ☐         what you did ☐

your results ☐         your conclusion ☐

**Safety notes:** Don't hold the material between the light source and your eyes – you could hurt your eyes. Point the light away from you at the material. Watch how much light goes through it and onto a wall or other object.

## TASK CARD 2
# What can you make into a magnet?

*First find out: How can you use a magnet to make another magnet?*

**You will need:** materials to test, magnet, iron filings

**Things to think about:** Magnets are always made of metal. Make sure you include some different metal things to test. How will you know whether the things you test have become magnetic?

**Presentation ideas:** You could present your results as a science feature on a children's television programme.

**Have you included these things in your presentation?**

the question ☐         what you used ☐         what you did ☐

your results ☐         your conclusion ☐

## TASK CARD 3 — Which part of the tongue tastes salt?

**You will need:** volunteer tasters, salt, water

**Things to think about:** How many people will you test? What will you use to put the salty water on their tongues? How salty will you make it? (It must be the same saltiness for each person.) Which parts of the tongue will you test?

**Recording your results:** You could draw up a tongue for each person and mark on it where they taste salt. Or you could decide on names for the different parts of the tongue (front, back, middle, side) and write down which part tastes salt for each person.

**Presentation ideas:** You could draw a huge tongue and mark on it where each person could taste the salt best. You could use a different colour for each person.

**Have you included these things in your presentation?**

the question ☐     what you used ☐    what you did ☐

your results ☐    your conclusion ☐

**Safety notes:** Use clean things to measure out the salt and stir the salt into the water.

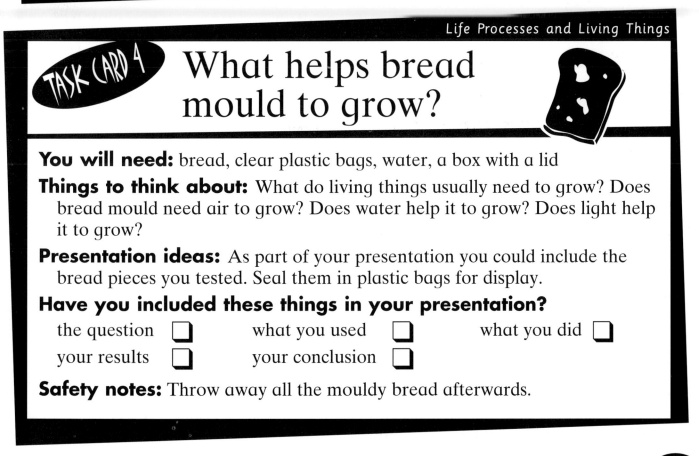

## TASK CARD 4 — What helps bread mould to grow?

**You will need:** bread, clear plastic bags, water, a box with a lid

**Things to think about:** What do living things usually need to grow? Does bread mould need air to grow? Does water help it to grow? Does light help it to grow?

**Presentation ideas:** As part of your presentation you could include the bread pieces you tested. Seal them in plastic bags for display.

**Have you included these things in your presentation?**

the question ☐    what you used ☐    what you did ☐

your results ☐    your conclusion ☐

**Safety notes:** Throw away all the mouldy bread afterwards.

## TASK CARD 5 — How far can you drop a water balloon without it bursting?

**You will need:** balloons, water, safe places to stand at different heights (outdoor steps are ideal)

**Things to think about:** What kind of balloons will you use? (They should all be the same size.) How many will you test? How full will you fill them? What heights will you drop them from? How will you measure (or estimate) how high you are? How will you stay dry? Will you need to clean the water up afterwards?

**Presentation ideas:** You could present the results of this experiment as a demonstration. First explain to the class how you conducted your test, and why you decided to do it that way. Then show what happens when you drop the water balloons from different heights. Make sure your audience stands well back!

**Have you included these things in your presentation?**

the question ☐        what you used ☐        what you did ☐

your results ☐        your conclusion ☐

**Safety notes:** Make sure you are standing safely before dropping the water balloons.

## TASK CARD 6 — How much juice is in an orange?

**You will need:** oranges of the same size (or as close as possible)

**Things to think about:** How will you get the juice out of the orange? How will you make sure no juice is lost? How will you measure how much juice there is?

**Presentation ideas:** You could use a bar chart to show how much juice was in each orange. You could also work out the average amount of juice in your oranges. To do this, add up the total millilitres of juice, then divide it by the number of oranges you used.

**Have you included these things in your presentation?**

the question ☐        what you used ☐        what you did ☐

your results ☐        your conclusion ☐

**Safety notes:** Ask your teacher for help if you need to cut the oranges.